Introduction

The object of this book is to take the reader on a photographic trip down Memory Lane to look at working commercial vehicles covering approximately the 20-year period from 1963 to the early 1980s, although most of the content dates from the early 1960s.

We all have a favourite part of our lives that we look back on with pleasure and nostalgia, particularly where transport memories are concerned. Just what that time is probably depends upon the period in which each individual reader grew up — rather like a rainbow, everybody has their own personal vision. Certainly for most of us our formative, childhood years are the ones that have the most bearing on our outlook in later life. It will, therefore, be a rolling date, to be brought forward by each subsequent generation of transport enthusiasts and in, say, 30 years' time there will be those who look back on the 1990s as their own preferred time.

The writer grew up in the late 1940s/early 1950s in Nottingham, in an area close to the Forest Recreation Ground, a large open area where the mammoth Goose Fair is held annually in October, and where many of these illustrations were taken. Early autumn Monday mornings, before going to school, were spent watching the queue of showmen's vehicles moving on for the build-up, and all my spare time out of school was spent at the fair. These early experiences presumably gave me the love of showmen's transport that I still possess, and is possibly reflected in the proportion of the book given over to this type of transport. I hope the reader will bear with me in this indulgence; certainly there are many, both transport enthusiasts and modellers alike, who share this interest, and for those who are not so afflicted, I hope that the variety of vehicle makes and types will more than compensate for their inclusion.

In the early 1960s I was fortunate to be able to buy a 35mm camera that took colour slide film, and all the illustrations were taken by me with that camera, which is still in use. By today's standards both the camera and the film used hardly bear comparison, and at times this is reflected in the quality of reproduction. Please bear in mind, however, that many of the slides used are well over 35 years old.

It should also be stressed that the vehicles were not posed, but photographed 'as found'; at work, parked up for the night, or, in some instances, in scrapyards. Sometimes the camera had to be pointed at the subject and the shutter quickly released to avoid missing a particular vehicle, possibly forever. There was often no chance to wait for improved or perfect lighting, nor to wait for a person or other vehicle to move to improve the make-up of the picture. I trust that the quality of the subject matter more than outweighs any photographic shortcomings.

Whilst some vehicles are obviously well kept, the majority do not have the rally 'spit and polish' appearance so evident in many photographic works today. This is not to decry these books — indeed they provide useful reference works in their own right — but there is no denying that the subjects are artificially posed in perfect weather and lighting conditions, and the real character and feel of the period represented is missing.

As far as is known, at the time that the photographs were taken all the vehicles illustrated within these pages were working vehicles, none being in preservation. Sometimes dented and scraped, sometimes rusty and grimy and with faded paintwork, they nonetheless had a dignified and workmanlike quality.

The classic long-distance trunker type of vehicle has been well recorded by the likes of Peter Davies, but what have not been so well recorded, perhaps

First published 2000

ISBN 0 7110 2704 8

Published by Ian Allan Publishing

an imprint of Ian Allan Publishing Ltd, Terminal House, Shepperton, Surrey TW17 8AS.

Printed by Ian Allan Printing Ltd, Riverdene Business Park, Hersham, Surrey KT12 4RG.

Code: 0002/B2

Title page: In the 1920s Harrods, the famous Knightsbridge store, was using a fleet of solid-tyred, American-built, battery-electric vans for its parcel delivery service. By the mid-1930s these were showing their age, and there was talk of replacing them with petrol-engined vehicles. Harrods' Chief Electrical engineer, Mr J. H. L. Bridge, was, however, a strong believer in battery-electrics, and was convinced of their suitability for door-to-door delivery in the classier inner suburbs where Harrods' customers tended to live.

Not being satisfied with what was commercially available at that time, Mr Bridge designed a 1-ton payload van using a nickel-iron battery that was lighter than a conventional lead one (the all-up weight including the driver and load was nearly 3 tons, an advantage over existing production vehicles). Power was taken from the battery to an underfloor-mounted electric motor which transmitted drive to a modified Morris truck differential via a propshaft. It was claimed that the design had an efficiency of 82-3%.

The prototype appeared and was approved in 1938, and a further 60 were built by Mr Bridge and his staff in Harrods' underground garage workshop at an average of one van every two weeks, the fleet being completed in 1941. This example, FLC 520, was registered in November 1938 and was in virtually showroom condition when seen in a South Kensington street on 6 May 1963.

Right: Pictured on 8 August 1963 is a classic 1940s example of a turntable-ladder fire appliance by Merryweather. GTO 10 was operated by the City of Nottingham brigade, and was photographed within the confines of the central fire station in the city centre. The massive ladder has supports on either side of the handsome radiator to secure it when in the travelling position, and an impressive searchlight can be seen mounted on the offside.

To see one of these vehicles in full flight on its way to a fire was a splendid experience — engine at full throttle, bell ringing and crew hanging on — and cannot be recreated by present-day machines, which look almost clinically efficient but lack that certain character.

Just glimpsed behind is part of a motorcycle combination, one of several that were used for hydrant testing, whilst to the left is a period petrol pump.

due to their more mundane image, are those vans, fire engines, tower wagons, breakdown lorries and tankers that carried out a wide range of everyday tasks, keeping the wheels of our lives running behind the scenes. Neither should we forget the showmen's transport that brought the circus and fair for our leisure time enjoyment.

Unfortunately, there was a tendency to overlook some vehicles on the basis that they could always be photographed later. Indeed, to me some vehicles gave the impression that they would go on forever, and at the Goose Fair some old favourites could be seen year after year. However, all this was to change, for the early 1960s saw the end of the austerity era that had lasted from the end of the war in 1945. Image began to be of importance to fleet operators, more money was available to buy from a growing range of vehicles, and increasing legislation loomed on the horizon. This was to bring to an end the 'make do and mend' era which had kept many vehicles soldiering on in a second life, converted into specialist service vehicles. In future such vehicles would be specially built on new chassis.

Whilst studying the illustrations, the reader is invited to take note of examples of vehicle lighting that would now be illegal, the general lack of flashing trafficators, the small cab mirrors and, in several cases, the complete lack of cab side window glazing. All these factors, coupled with crash gearboxes, heavy clutches, no power steering and usually no heater or demister would not be tolerated by today's professional drivers. That said, traffic volumes were much lighter than they are today, parking was not the nightmare it is now, roadsides were less cluttered with the plethora of signs that are apparently deemed so necessary today, and just look at the absence of yellow lines!

Throughout the book, I have used the word 'lorry' for British-manufactured vehicles, whilst 'truck' is used for British-built light pick-ups or American-produced vehicles. Any reference to war does, of course, refer to the 1939-45 conflict.

Dates and details are taken from my own records; if any information is incorrect or incomplete then I apologise, but would add that, in the early 1960s, there was less information available for the enthusiast to refer to, and much of it could only be obtained by dint of detective work.

It all seems so far away now and I hope you, the reader, will enjoy these photos as much as I enjoyed selecting them. If it provides some nostalgic pleasure to those who remember that period, and if those too young to have known it learn a little of what it was like, then the effort will have been worthwhile.

Barry S. Watson ACIB

Left: In the early 1930s the city of Nottingham had a large tram fleet and an expanding trolleybus system. There arose a need for a reliable tower wagon, and in 1933 the decision was taken to convert AEC Regent No 48, a Park Royal-bodied machine that had been new the previous year, for this rôle. Its body was removed, it was converted to normal-control layout and an Eagle tower was fitted. The vehicle ran in this guise until the abandonment of the trolleybus system in 1966.

It was photographed standing by the trolleybus stop, adjacent to the time clock in the Old Market Square, on 7 May 1964 whilst a member of the crew had a discussion with a transport department inspector. It is interesting to note the width markers mounted on the mudguards, the scuttle-mounted sidelights, and also that only one original headlight remains. No flashing

warning lights are fitted to the vehicle, although the 'overhead repairs' sign on the roof does have a small lamp to illuminate it for night-time work.

Above: The AEC Mammoth Major six-wheeler was introduced in September 1931, and the model name was one that was to last for many years.

This early 1930s example, registered in Burnley as HG 1014, is seen parked in the company of some period cars and a 1956 Morris 1/2-ton van, at a small fair set in rural surroundings at Thrybergh, near Rotherham, on 1 September 1963.

This very original machine was owned by John Pullen of Rotherham, and sported a beautifully varnished, bow-roofed, box-van body that was subsequently transferred to a more modern chassis.

Above: To house its expanding fleet, Nottingham City Transport opened a new depot and offices at Carter Gate, subsequently renamed Parliament Street, in December 1928.

Standing outside the depot on 26 May 1963 is DAU 455, a 1937 AEC Regal staff canteen. Originally numbered 76, this vehicle was renumbered in 1944 and again in 1948 before being converted for this rôle, as 812, in 1956. The 32-seat Cravens body was little altered, although a large water tank was mounted on the front portion of the roof. It continued in use until 1968.

The outline of the former tram tracks, tarmac-covered, can easily be seen in the roadway, whilst just visible within the depot is an 8ft wide, Brush-bodied, six-wheel BUT trolleybus.

Right: By 1950 Nottingham had an extensive trolleybus network, with a fleet of approximately 150 vehicles, and the need arose for a new tower wagon. The decision was taken to convert a 1939 AEC Regent bus, fitted with a Metro-Cammell body, and 31, FTO 614, was chosen, becoming service vehicle 802. It is seen parked in Parliament Street, near the top of Queen Street, on 5 April 1964, whilst the line crew attend to some overhead maintenance.

When viewing this scene from the standpoint of the safety-conscious 1990s, it seems strange to note the total absence of flashing warning lights, marker cones and high-visibility jackets. All this was, of course, considered unnecessary at a time when traffic was so much lighter and speeds lower. Also of interest, parked at the roadside, is the Austin Mini Countryman, nearly new when the photograph was taken.

Left: The AEC Matador 0853 was developed from a design by FWD, originally using a Hardy Motors 4x4 chassis that used AEC components, before Hardy was absorbed into the AEC empire. With its AEC six-cylinder, 7.58-litre diesel engine, it became one of the most successful vehicles of the war, with well over 9,000 being built in quantity production that lasted from 1940 until 5 November 1945, when the last vehicle left the assembly line. The majority were built as artillery tractors, but they were also used as general service vehicles, and the RAF used them for trailer haulage.

With the return of peace the off-road ability of the design was quickly recognised by showmen, and this 1948-registered example, LUM 955, became well-known hauling Harry Lee's famous Steam Yachts ride. One load for the Matador would comprise the two yachts, *Shamrock* and *Columbia*, each on its own trailer, plus the trailer-mounted steam engine that provided the motive power for the ride. As used by Harry Lee, the

body was fitted with a generator and a demountable crane jib, clearly visible in the photograph, for loading and unloading the heavy yachts.

Attending a traction engine rally on the outskirts of Stamford, Lincolnshire, in June 1964 the AEC is flanked by a Commer Superpoise on the left and a Guy van on the right.

Above: Henry J. Coles built his first crane in 1879, and the Coles name became one of the best-known in this specialist field. When road vehicle-mounted cranes came into use, a solid, well-built chassis was a prerequisite, and the AEC six-wheeler seen here fulfilled that role admirably.

Standing on the harbour wall at Ramsey, Isle of Man, on 20 May 1975 is NMN 260, a lattice-jibbed example, registered in 1951 and owned by the harbour board. Smaller ports or dock authorities would not need to invest capital in rail-mounted or permanently sited cranes, and the flexibility provided by such a road vehicle would meet all their needs.

Above: Regarded by many as one of the all-time classic designs of the British commercial vehicle scene is the eight-wheel AEC Mammoth Major V, with a cab that was built by Park Royal, not far from AEC's Southall works.

The photograph shows 9028 JH, a 1960 example owned by Southampton-based Stokes Fairs, at Swanage, Dorset on 22 August 1980. The body is a typical double-deck showman's frame type, with additional storage for small items provided by the box mounted on the rear nearside of the chassis. The vehicle is backed up to the Speedway ride that it has brought to the fair, and to the left can be seen a set of Chair-o-Planes.

Right: In the early 1960s Chipperfield's had a very large touring circus which required a sizeable transport fleet for moving its equipment and animals from town to town.

With the circus in Nottingham on 4 April 1964, parked behind part of the menagerie, is this 1936 Stoke-on-Trent-registered, six-wheel Albion, with simple framework mounted on the original flat platform body. It is painted in a basic fashion, with no lining-out or lettering. The whitewall tyres provide a nice period touch, being in vogue on many cars at that time.

Above: During the war there was a great need for heavy tractors, and Albion was asked to design a vehicle to complement the Army's Scammell Pioneer. The result was the CX22S, a 6x4 model introduced in 1944 and only produced in relatively small numbers.

The example shown here, bearing a 1946 London registration, was owned by Holland's Amusements, and is seen pulling on to the Forest in Nottingham for the Goose Fair on 30 September 1963. The double-skinned cab roof, demonstrating its military ancestry, can be clearly seen. It is interesting to note that the trailer wears a completely different colour scheme, although the Albion was to keep its green paint for many years.

In the background can just be discerned a prewar Leyland and a wartime Crossley.

Right: Found in a Staffordshire scrapyard on 12 April 1981 was this postwar Albion 6x4, a former army mobile workshop unit. This 10-ton vehicle, designated FV11102, Albion, WD/HD/23N, was produced in the early 1950s, and was fitted with a six-cylinder, 160bhp petrol engine. The workshop body, built by Strachans (Successors) Ltd, was insulated and could be warmed or cooled according to the climate in which the machine was operating.

During the 1950s and '60s the heavy, rigid, eight-wheel design was Atkinson's main product, and was produced with little change until after the takeover by Seddon in 1970.

The basic cab design as illustrated was introduced in 1958 and was made of reinforced glass fibre mounted on a jig-built hardwood frame. Comberhill of Wakefield, West Yorkshire, was for many years the Atkinson distributor for Yorkshire. The firm's Cummins-badged machine has a very workmanlike air, and makes a fine sight parked on the outskirts of Wakefield on 12 July 1975.

At one time it was customary for small garages to adapt car chassis to make pick-up trucks, or small breakdown vehicles, and this 'make do and mend' system continued well after the austerity years following the war, until such time as new, affordable replacements came onto the market.

Many of the earlier conversions soldiered on for several years, especially in rural areas, and TK 690, a 1928 Austin 12/4, was found at Holbeach,

Lincolnshire on 4 September 1965. The 12/4 was a sturdy, long-lived chassis that formed the basis of many of London's taxis in the 1920s.

Many period items can be seen in this view; the water temperature gauge on the radiator, the scuttle-mounted sidelights, the running-board-mounted battery, the caravan, a 1930s Vauxhall Light Six and, in the background, a Power petrol pump and a wall-mounted National Benzole advertising sign.

Providing a welcome break for travellers, in a lay-by on the Derby to Uttoxeter road, was MJ 8302, a former ambulance converted into a mobile tea bar. The chassis was an 18hp, six-cylinder Austin, believed to date from about 1933, and based at Mickleover, near Derby. It was photographed on 19 April 1964, a damp and misty day, in company with a mid-1950s Vauxhall Velox or Wyvern.

The cattle market in Nottingham was a regular parking spot for lorries at weekends or overnight. Seen here on the cobbled road surface on 25 April 1963 is a 3-ton, 6x4 Austin K6 airfield fire engine, en route from a disposal sale at Ruddington. A trade plate is fastened to the front bumper, and its bell is still in place. It was registered HXA 873 (1946) and was probably still fitted with its original six-cylinder, 3.99-litre petrol engine.

Above: Photographed in the grounds of Staunton Harold Hall, on the borders of Nottinghamshire and Derbyshire, on 31 May 1964 is SUT 150, a 1959 Austin Gipsy fire engine, seen with its front-mounted pump connected to a water supply.

The Gipsy was Austin's answer to the successful Land Rover and was introduced in 1957. This particular model had all-round independent suspension at a time when the Land Rover was leaf-sprung, but leaf springs were offered as an option on the Gipsy from 1962. In due course Austin, by virtue of being part of British Motor Holdings, joined the Leyland Motor Corporation, which already owned Rover, and the decision was taken to drop the Gipsy in favour of the continued production of the Land Rover.

The example illustrated was owned by the East Midlands Division of the National Coal Board and was based at Coleorton.

Right: Environmental pressures may well lead to the disappearance of those scrapyards where vehicles were allowed to moulder over many years. At the end of their working lives most commercial vehicles ended up in such places and this is what has happened to this trio of sad Bedfords, awaiting their fate in Staffordshire, on 12 April 1981. On the left is a J1 model van and next to it is an early model OY 3-tonner. At the far right is a 4x4 RL dating from 1962 and fitted with an Atkinsons of Clitheroe gritter body.

Bedford was a famous marque and its loss is still mourned by operators, drivers and enthusiasts alike.

ATT 921 is a Bristol J that was new to Western National in 1935. After the war there was an acute shortage of new chassis, and its well-worn original timber - framed body was replaced by a new one, manufactured by Beadle of Dartford.

Subsequently sold to a showman, the vehicle was found at a traction engine rally near Stamford, Lincolnshire, in June 1964. With its Gardner engine used for generating power the bonnet side panel has been removed to assist with cooling and a drum of diesel fuel lies by its side. Boards have been added to the roof to enable it to be used for carrying extra equipment.

During the war Canada became a main supplier of vehicles to countries within the British Commonwealth, and a whole range of types was designed, becoming known as the Canadian Military Pattern (CMP).

The standardised designs were produced by General Motors, Ford and Chrysler, and here is JTV 168, a 1947-registered Chevrolet C15 truck, in use with a jobbing builder/joiner in Nottingham on 22 September 1964. It will be noted that the cab is fitted with a plastic sidescreen, instead of a glass window, and it boasts semaphore trafficators.

The road sign on the left is of a type now long discontinued, and there is an interesting array of popular cars of the era.

During the war there was a need for the road movement of complete fighter aircraft and a semi-trailer for this purpose was designed, developed and built exclusively by Taskers of Andover. The popular name for these was 'Queen Marys', after the Cunard liner of the same name.

One of the tractors that were used to pull these trailers was the Commer Q2, an 8ft-wheelbase vehicle, powered by a six-cylinder, 4-litre petrol engine. An example is shown here in its later, civilian guise as a mobile welding unit. It was used for site repairs to plant equipment by its owner, Cripps of Nottingham, and is seen in the firm's yard on 1 September 1964.

A range of light utility vehicles for use by the armed services was produced by all the major manufacturers during World War 2. The Rootes Group contributed a truck based on the 92in-wheelbase Hillman Minx of 1939. After the war this was developed into a Commer light van using the updated bonnet and grille from the then current Minx car.

This tidy example, dating from 1946/7, was parked at Laxey, on the Isle of Man, on 21 May 1975.

The early postwar 25cwt Commer was more usually seen in van form, and was advertised by the Rootes Group as being 'Ideal for the localised delivery of loads which are bulky yet relatively light'. Designated the Q25, the model was fitted with the Rootes 40bhp four-cylinder petrol engine.

EFE 665 is a very original 1950 small lorry version, owned by the Boston Diesel Engine Co Ltd, and photographed at the Showground in Newark, Nottinghamshire, on 9 May 1964.

The gently-shelving sandy beaches of Blackpool present the pleasure boat operators of this Lancashire resort with the problem of how to transport trippers to the boats whilst keeping their feet dry.

In this instance the problem has been solved by the use of two ex-Army Commer Q4 4x4 lorries which, when put into service, had their cabs converted to open-top layout. These dated from the 1950s, and sported 95bhp six-cylinder petrol engines. They were well-patronised when seen on a warm and sunny 11 October 1975.

Photographed in a Langley Mill scrapyard, on the borders of Nottinghamshire and Derbyshire, on 5 September 1963, is FA 1075, a 1921 Dennis fire engine. This Burton-on-Trent-registered machine was delivered new to Bass, Ratcliffe & Gretton Ltd, the brewers, and had just been withdrawn with a mileometer reading of less than 17,000 miles. At some time in its history the original solid tyres have been replaced by pneumatic tyres, but the original headlamps and bulb horn remain, as does the 'Fire Brigade No 1' legend on the faded paintwork of the Braidwood body. A marvellous assortment of withdrawn commercial vehicles surrounds this historic gem, which has since been preserved.

Introduced at the Olympia exhibition in 1931, the Dennis Lancet I single-deck bus chassis soon found ready buyers, and its sales helped to keep Dennis busy throughout the Depression years.

Touring with Chipperfields Circus in 1964 was BAA 390, a 1936 model that was new as Aldershot & District 677. This clean vehicle, its original Strachans body replaced by this purely functional one, is here seen doubling as an anchorage point for the adjacent tent side wall, on the Forest at Nottingham on 4 April 1964.

Left: Dodge commercial vehicles were built at Mortlake Road, Kew, Surrey from 1931 but, despite having the backing of the parent company, Chrysler, never managed to achieve the sales levels of Ford or Bedford (General Motors).

GXM 552 is a 1943 Dodge pump water tender of the Hampshire Fire Brigade, a handsome rebuild of a former wartime water tender. Parked at the rear of the fire station at Lymington, in Hampshire, on 2 June 1963 it is in the company of some typical cars of that era, including a 1930s Vauxhall. At that time it was still not uncommon to see prewar cars in daily use.

Above: EAP 369 is a 1948 Chrysler-designed British Dodge, fitted with a Perkins P6 engine. The articulated version of this model was never common, and this horsebox, complete with horse bonnet mascot, from East Grinstead in Sussex, is very unusual. This cab style was to last for only one more year before the introduction of the Briggs-designed cab with the ridged bonnet.

The Vauxhall Velox dates from 1961, whilst the Bedford CA is a 1958 model. The photograph was taken at the Newark Agricultural Show, Nottinghamshire, on 9 May 1964.

Above: The 4x4 American Dodge weapons carrier with winch was highly regarded as an infantry vehicle in the war, and many were supplied to the Allies under the Lend-Lease programme. They were very durable, and large numbers continued in use in various guises all round the world after the cessation of hostilities.

This particular example, converted to a breakdown truck, was seen in a yard at Wells-next-the-Sea, Norfolk, on 5 September 1965, keeping company with several decaying cars.

Right: In the late 1950s the Leyland group and Dodge were both in need of a new cab, and a decision was reached whereby Motor Panels would produce a standard cab for them, development costs being shared by Leyland and Dodge. The cab was exhibited on the Motor Panels stand at the 1958 Commercial Motor Show, and was known as the LAD (**L**eyland **A**lbion **D**odge).

Dodge has never been a common make in showland use, and 352 BBD, with an LAD cab, was considered a rarity when photographed at the build-up of Nottingham Goose Fair on 27 September 1981. It was registered in Northamptonshire in 1962.

Above: One of the most respected chassis for heavy goods haulage for many years has been the ERF eight-wheeler. Having proved to be reliable, strong and durable, many find their way into a period of 'after-life' in showland, where they continue to earn their keep the hard way.

ENT 575 is a 1948 example, and must have been one of the last produced before the introduction of the streamlined cab. It is seen with Raymond Armstrong's Waltzer at Burghley Park, Lincolnshire, in June 1964. Armstrong's was renowned for the boldness of its lettering and in this case has used the lengthened body to full advantage. Despite its hard life, the vehicle is clean and tidy, and is obviously well looked after.

Right: When the KV (Kleer Vue) style of cab was introduced by ERF at the 1954 Commercial Motor Show it attracted considerable attention. With its wrap-around windscreens and nonconformist radiator grille shape, it was a clean break with tradition. Built of glass fibre, on a framework of wood and steel, by Jennings of Sandbach, the screen curvature must have tested the skills of the development engineers at Triplex, which supplied the glass. The cab had the advantage of light weight, and provided first-rate visibility for the driver. When first introduced it had single headlamp units.

This fine example, 2028 EH, seen at Nottingham's Goose Fair on 9 October 1982, was owned by George Godden, travelling with his Twist ride, and is a later production model with twin headlamps.

Above: When seen at a Staffordshire traction engine rally, this July 1938-registered Foden DG articulated tractor unit was owned by the North Staffordshire & Cheshire Traction Engine Club. It had been used to take a steam exhibit to the rally.

This was a comparatively early production model, and would be most welcome at a commercial vehicle rally today, but when seen on 21 July 1963 it was relegated to the perimeter of the rally site, being regarded with less favour than its steam-powered load. The semi-trailer is of the knock-out rear axle type, and has been left with the wheels removed to save time when the steamer is ready to be reloaded.

Above right: Forming part of Chipperfield's sizeable circus transport fleet is this Foden DG6. This is an ex-Army model, featuring a stylishly-lettered body which was custom-made for its new role. It was attending the circus at Nottingham on 4 April 1964.

As developed for War Office use, the DG6 was designed to a gross weight of 19 tons, and 1,750 were supplied from 1940 to 1944. They were fitted from

new with a Gardner 6LW, developing 102bhp, which drove through a five-speed constant-mesh gearbox that included a crawler gear for difficult conditions. This feature, together with the Gardner's legendary economy and reliability, readily endeared them to showmen and circus proprietors in the early postwar years.

Right: This ex-army Foden DG6 is the same model as the Chipperfield's vehicle shown previously. The reason for including both is to illustrate the completely different treatment that the two vehicles have received. This one has received a simple flat platform body, lacking stylish front mudguards, and the wartime track grip 13.50 x 20 tyres have been replaced by those of a smaller size which don't sit too happily under the simple pattern mudguard. It is not known whether the rear mudguards were specially fabricated, but they bear a strong resemblance to those fitted to the six- and eight-wheel Sentinel S-type steam 'waggons' of the 1930s. Registered 890 SRE, it was seen at Wistaston, Cheshire on 6 September 1964.

Left: The DG Foden was renowned for its strength and longevity, and this DG6 was 35 years old when seen at Harrogate Stray Fair on 25 August 1975. Owned and well maintained for many years by its Leeds-based showland owner, it has received modifications such as the FG-type front bumper and a more recent chrome trim addition to the radiator grille. It still has its original headlights and starting handle, and all the paintwork and trim gleam.

Above: To many enthusiasts the classic British heavy goods vehicle is the rigid eight-wheeler, and the genre is typified by GOA 652, a 1945 example of the Foden DG. This gem was owned by Tower Hill Transport, the trading name of Boston Stevedores Ltd, an organisation formed by two ex-BRS depot managers in 1955, the Tower Hill name being used for the long-distance fleet.

Loaded with round steel bars, the vehicle is here seen parked up for the night in a side road near Trent Bridge cricket ground, Nottingham, on 2 September 1964. The complete absence of parked cars and yellow lines is in stark contrast to the typical 1990s urban road scene.

Above: Three examples of Foden's late 1940s production types were captured together, attending a fair at the Market Bosworth traction engine rally in Leicestershire on 23 August 1964. All are clean, tidy and unlettered. From left to right are: FNT 132, a 1949 FG; EJF 796, a 1947 DG; and FAW 927, a 1949 FG which clearly shows the early styling of the S18 cab.

The vehicles were used to transport a set of Dodgems, and one of these cars can be seen on the upper level of the body, above the cab, of FNT 132.

Right: Charles Thurston's Amusements of Norwich was always renowned for its well-turned-out transport fleet, and this early example of a Foden FG6 is typical. KGT 979, dating from 1949, is clean, well-painted and lettered and is an excellent advertisement for its owner's Hurricane Jets ride. Interestingly, it sports the type of bumper that was fitted to the later S20-cabbed models. It is parked at the edge of the Forest, in Nottingham, on 4 October 1964, coupled to the trailer carrying the air compressors for the ride.

The Fordson 51 model was never the most common of Ford's 1930s commercial vehicle range. Seen here is a 1936 example, complete with a varnished wood horsebox body which is cut off very short behind the rear wheels. A drop-down loading ramp is incorporated in the offside of the bodywork, whilst rails are fitted to the cab roof for the carriage of hay bales.

RD 8974 was still giving good service to its lady driver in its original guise some 28 years later, on 9 May 1964, at the Newark, Nottinghamshire, Agricultural Show.

GLW 364 is a 1942 registration, and this Fordson 7V was in use with the local division of the Civil Defence Corps when seen in Peterborough on 20 June 1964. With its Ford family grille and large headlights it made a fine sight, particularly as it was in such good condition.

A period Hillman Minx convertible can be seen parked in the background.

At one time every rural district had its own carrier for the delivery of small items in the locality. Bullers of East Bridgford, a small village some miles east of Nottingham, used the 'Carrier' title on JAL 507, a 1947 Fordson 7V.

The 7V was a postwar model, carried over from one that was introduced in 1937. It could be fitted with either a four-cylinder engine or, more usually, a 30hp, side-valve V8 petrol engine of 3,622 cc. The gear-change layout in these vehicles was awkward, the driver having to reach behind him to change gear and risk banging his elbow on the back of the cab. Ventilation to the cramped cab was provided by front dash-mounted vents and a sliding roof.

JAL 507 wears the final style of slatted radiator grille and has the smaller headlights that were sometimes fitted to this model. The dropside body is a standard Ford product. It was photographed at Broad Marsh, Nottingham, on 27 April 1963. Behind can be seen the old Great Central Railway viaduct and bridge carrying the line between Nottingham (Victoria) and London (Marylebone).

The successor to the Fordson 7V range was the ET6, introduced in prototype form at the 1948 Commercial Motor Show, with production starting in April 1949. Originally offered in the 2-5-ton payload range, there was also a six-wheel variant known as the Sussex.

The 30hp Ford side-valve V8 petrol engine remained a standard fitting, but the Perkins P6 was an option, whilst further choice was provided later by Ford's famous 'Cost Cutter' four-cylinder, over-head valve (OHV) petrol unit. Briggs Motor Bodies built the cab, considered an improvement over the 7V cab as there was no bulky intrusion by the engine cowling.

LYR 240 is a 1951 version in the form of a police mobile control room. The mast on the nearside mudguard is to assist the driver in positioning the vehicle when manoeuvring in tight spaces. It is seen at the Lincolnshire Showground on 19 August 1978.

Left: Announced in March 1957 as a new medium lorry range, the Ford Thames Trader soon became a regular sight on Britain's roads, easily recognised by its short 'snout' bonnet.

409 AUU is a 1960 example, chassis number 510E63308, fitted with a six-cylinder diesel engine of 5,416 cc. It was originally placed in service by London Transport as an articulated tractor unit, numbered 1191F, in June 1961 for use with a York semi-trailer. It was withdrawn in May 1969. It is seen five years later, on 28 July 1974, at a traction engine rally and carnival at Kegworth, Derbyshire, in converted form as a short-wheelbase drawbar tractor, and had been equipped with a small generator unit.

Above: GMC's amphibious vehicle contribution to the war effort was the DUKW-353 (D — 1942; U — amphibious; K — all-wheel-drive; W — tandem rear axles). Given this designation and its 'swimming' ability, it naturally became known as the 'Duck', and more than 21,000 were built.

A metal boat-type hull was fitted to the 2¹/2-ton 6x6 CCKW chassis, fitted with a straight-six, 4.4-litre petrol engine. The body could accommodate 25 men, and these vehicles were used to good effect in many theatres of the war, landing troops and supplies from ship to shore. When travelling through water the propeller was driven from the bottom three gears of the five-speed gearbox, giving a speed of between 3 and 5 knots. The driver had the facility to vary the air pressure in the tyres from his driving position, this being particularly useful when changing over from running on hard surfaces to soft sand.

This example on Southport beach, on 6 August 1976, was being used by Sefton Metropolitan Borough Council for beach patrol duties, and appears to have been modified only by the addition of a ladder to make for easier access.

Left: At one time fairgrounds regularly yielded the sight of retired buses, working out their last days as haulage vehicles, and sometimes caravans, for their showland owners. With their low-slung chassis and long rear overhangs, they would not appear to be suitable for the uses to which they were put, and yet they seemed to cope with muddy and rough sites with remarkable aplomb, scraping over uneven ground and long grass.

HRA 682 was a 1942 Guy Arab, fitted with a Brush utility body and a Gardner 5LW engine, owned by Hall's Amusements and photographed in Derby on 31 August 1963. It was once number 187 in the Midland General fleet, and had always been well maintained in that operator's smart blue and cream livery. Little apparent external work was put into converting it to its new rôle, although the passenger windows were painted over for privacy and to conceal its load. In this guise it ran for a number of years in the Nottinghamshire/Derbyshire area.

Guy is best remembered for its Arab passenger chassis and, in the latter years of its existence, for the Big J lorry chassis, but at one time the lightweight Wolf and Vixen chassis sold in respectable numbers.

This early postwar Vixen may have the wrong model name on its radiator grille as it was usually only offered as a forward-control model, the Wolf usually being the model that was offered as a bonneted option. Fitted with a neat breakdown body and crane, and with its headlamps faired into its wings, it is seen at a garage on the Leeds ring-road on 1 August 1976. It is believed to have been owned at one time by the Leeds Co-op Society.

48

Left: The final model of the Guy Otter diesel was fitted with a version of the glass fibre cab that Guy had introduced at the 1958 Commercial Motor Show. Being described by *Commercial Motor* as 'outstanding in appearance', the cab was certainly considered revolutionary at that time, having a large wrap-around windscreen with a projecting cab visor and dual headlamp units.

Whilst the cab sat well on the heavier chassis it did not look so well-suited to the lighter, Gardner 4LK-powered versions, one of which is seen here at the Lincolnshire Showground on 19 August 1978. Registered CUB 48C, it dates from 1965.

Below: One of the postwar designs to replace World War 2 vehicles was the Humber 1-ton, 4x4 truck, built by the Rootes Group in Maidstone and known as the FV1600 series. Powered by a Rolls-Royce B60 six-cylinder petrol engine of 4,250 cc, and equipped with a five-speed gearbox and all-round independent suspension, it had a very good cross-country performance.

The very smart and virtually standard vehicle seen here coupled to a drawbar trailer was at the Newark Show, Nottinghamshire, on 9 May 1964. 335 VRR had only been first registered for civilian use in January of that year.

Left: During the 1930s the Americans had been working on perfecting a half-track military vehicle, for use both as an armoured personnel carrier and as a self-propelled weapons carrier, using the French Kegresse system with tracks incorporating rubber inserts.

The final design, which was a development of the American scout car, was perfected towards the end of the decade. Over 40,000 were produced by Autocar, Diamond T, International Harvester and White, many being supplied to the Allied armies under the Lend-Lease programme.

This example, built by International Harvester, was seen at Mundesley in Norfolk on 9 September 1965, being used to move beach huts off the beach by way of a slipway.

Above: Leyland Tiger AG 8285 was supplied new to a bus and coach operator in Ayrshire in the 1930s. For all its years on the road it still looked to be in sound condition when photographed on 16 August 1964 in Basford, a suburb of Nottingham, whilst working for a showman.

The vehicle had been used for generating electrical power during the fair and to do this the propshaft was disconnected from the rear wheels and the engine was run in gear with a power take-off turning a dynamo. By the time of the photograph, the fair was over and the transmission was in the process of being reconnected. The crude but effective levelling at the front of the chassis, by means of pieces of timber placed under the front wheels, will be noted.

Left: Leyland's early 1930s Beaver model, fitted with the 8.6-litre, indirect injection, overhead cam (OHC) six-cylinder diesel engine was a strong and reliable chassis which, with its Badger, Hippo and Octopus stablemates, helped Leyland to boost sales dramatically. The basic engine design, with only detail improvements, continued until after the war.

This example, AGY 841, dating from 1933, seen during the build-up of Nottingham Goose Fair on 29 September 1964, looks very sound for its years, and carries a typical prewar-style wooden showman's van body. The vehicle looks to be very original although the headlights appear to have been changed. The set of galloping horses just visible behind it was steam-driven, and has since been taken into active preservation.

Right: Loughborough Fair is held in the streets of the Leicestershire town every November. Parked there, on 16 November 1963, was this prewar, Northamptonshire-registered Leyland Beaver, NV 8250, originally operated by PX Ltd of Rushden. The tiny original sidelights and cab door mirrors will be noted, and the fact that there are no semaphore or flashing trafficators, hand-signals still being in regular use in the early 1960s.

In the early 1960s Rentokil Laboratories was operating a bird control service to prevent the damage done to large buildings by pigeon droppings. This involved the application of a special gel to the ledges on which the birds gathered. This gel made them feel insecure, persuading them to fly away and seek a roost elsewhere. Nottingham City Council was experiencing problems of this nature on its Council House, an impressive stone building in the Old Market Square opened in 1929.

Rentokil used ex-Fire Service turntable ladders, as, with an average reach of approximately 100ft, they enabled the crew to reach the lofty roosting sites. Seen here, on 30 June 1964, is DGJ 310, a 1936 London-registered Leyland with Metz turntable ladder. This handsome machine was fitted with a Leyland 8.8-litre, dual-ignition, OHC six-cylinder petrol engine. The first illustration shows the machine, levelled and jacked, with bonnet side panel removed to assist with cooling.

The second view shows the extent of the ladder's reach with the operator at the top making preparations to apply the gel. At that time the general public were free to pass close to the fire engine and walk under the ladder. Present day safety legislation would require the whole working area to be cordoned off.

In the early 1930s Leyland started to build a range of quality, lightweight machines to compete with the cheap, light and comfortable vehicles that were then being offered by Bedford, Ford and Morris. These were not built at Leyland's Lancashire base, but more than 200 miles away at the old Trojan works at Kingston-upon-Thames. The design, named the Cub in the Leyland animal family, was built in four- or six-wheel versions with either short or long wheelbase, and both goods and passenger-carrying versions were offered. Forward-control and bonneted versions were available.

Seen at Peterborough with a small touring circus on 22 June 1963 is ADP 136, a 1937 long-wheelbase passenger chassis Cub, model KPZ1, fitted with a Vincent horsebox body. At one time it may have been used by a wealthy racehorse owner, but when seen it had fallen on hard times, having broken its crankshaft and being towed between venues by the ex-Army Morris Commercial CS8 seen in the left of the picture. The elderly caravan and the Thames 400E van in the background would both now be considered collectors' items.

On 7 October 1963, DTB 152, a Lancashire-registered Leyland Metz turntable ladder, was being used to work on Blackpool Pleasure Beach's oldest ride, the 1904 Sir Hiram Maxim Flying Machine. This classic piece of equipment is in use to this day, and is a valuable historic relic in its own right.

The Leyland, supplied new to Morecambe & Heysham in 1938, is of interest in that it boasts a cab, giving the crew a degree of weather protection.

Up until about this time it was traditional for all fire engines to be fully open to the elements. Leyland's handsome, polished radiator fronting the gunmetal grey bonnet was an attractive feature for some years. The stabilising jacks can be seen in their wound-down positions either side of the rear wheels.

Above: By the early 1960s many towns and cities in the UK were actively scrapping their trolleybus systems. An exception was the Teesside system in Middlesbrough, where the splendidly-named Teesside Railless Traction Board was working on extending its route network.

On 14 June 1964 Leyland TS8 tower wagon BAJ 846 could be seen standing under the single overhead running wire that had so far been erected for the route extension at the Kingsley Road estate. This was to be the last extension made to any British trolleybus system.

Dating from 1939, BAJ 846 had a Willowbrook body, most of which had been retained in its original form forward of the tower. The Teesside system's attractive 'Maltese Cross' logo can clearly be seen both on the front headboard and on the side panels of the vehicle.

Right: EF 7352 was a well-used Leyland Lynx tower wagon, used by the Teesside Railless Traction Board for maintaining the overhead wiring of its trolleybus system, seen in the depot yard on 14 June 1964.

The semi-forward-control Lynx was a replacement for the Cub, introduced in 1937. It utilised many Cub components including the engine, fitted with an aluminium cylinder head to save weight and increase compression, enabling more power to be developed. This example carried a custom-built body which included an extra window above the nearside windscreen, so that a crew member could examine the state of the overhead wiring for faults as the vehicle was driven along. The sidelights are set back on the dash panel in vintage tradition.

Left: The 3-ton, 6x4 Leyland Retriever with a 5.9-litre four-cylinder engine was one of Leyland's staple products during the war, being fitted with a variety of bodies: general service, machinery, workshop and crane.

Many Retrievers were sold for further civilian use in the years immediately after the war when vehicles were in short supply and, inevitably, found their way into fairground use. MNU 224 is one of these; registered in Derbyshire in 1948 it carries quite a stylish body with valanced mudguards over the rear wheels and upswept rear panels to reduce the risk of grounding on uneven land. The curved front dome sits nicely over what would have been an open cab with folding canvas roof in the vehicle's service days. The drum-type headlights are unusual.

The vehicle was seen at Burghley Park, near Stamford, Lincolnshire, in June 1964, in company with a classic Atkinson which can just be seen in the background.

Below: Resting in a scrapyard on the Nottinghamshire/Derbyshire border on 23 May 1963 is GLW 412, an ex-Huddersfield fire engine. This World War 2 machine dates from 1942, when a small number of Leyland TD7 double-deck bus chassis were used as the basis for turntable ladders, to help meet the increasing demand for these appliances to combat fires caused by German bombing. When new it would have been delivered to the National Fire Service, painted in an overall grey livery. The typical half-cab driving arrangement of a bus was retained at the side of either a six-cylinder Leyland petrol engine, or an 8.6-litre diesel. The pump and ladder were both made by Merryweather, and the ladder could be extended to its maximum reach of 100ft in 30 seconds, power being taken from an engine-driven mechanical drive.

In the foreground is a 1957 Standard 6cwt van. This light commercial was based on the popular Standard 10 car, and used a 948 cc OHV engine. It was never to be seen in such large numbers as its competitors manufactured by Austin, Ford and Morris. Behind can be seen a variety of withdrawn BRS vehicles.

Left: The Mark II, 10-ton 6x4 Leyland Hippo, built for Army use, was introduced in 1944, and over 1,000 were produced, mainly intended for long-distance supplies haulage through Europe after the D-Day invasion.

The withdrawn example seen here has the standard GS cargo body, and was powered by a 7.4-litre diesel engine. The conning tower hatch in the cab roof can be clearly seen. It was photographed in a Staffordshire scrapyard on 12 April 1981.

Above: In the 1960s it was still possible to find custom conversions of vehicles for specialist uses, particularly if the operator was well-known for originality of thought and action. Barton Transport, of Chilwell, Nottinghamshire,

certainly fitted into this category, and at the time was one of the largest independent bus fleets in the country, operating an interesting assortment of vehicles.

Parked at the entrance to Chilwell depot on 31 August 1963 is service vehicle No 41, a converted Leyland. The low-slung chassis and the shape of the fuel tank indicate its PSV heritage, whilst a very smart, coach-built, full-fronted cab has been fitted.

Most of this originality of thought has gone nowadays, stifled by standardisation and the trend is for factory-built, 'off the peg' vehicles to be used, with the consequent drawback to the enthusiast of a lack of variety and interest.

Above: The replacement for the Leyland Lynx was the Comet, launched in December 1947, and becoming available for the home market in early 1948. Its manufacturers classed it as a 'cruiser-weight', and it shared its Briggs Motor Bodies cab design with Dodge.

This 1953 example, OGT 756, is believed to have been new to National Benzole as part of its tanker fleet, but when photographed on the outskirts of Wakefield on 17 September 1978 it was owned by Tarmac. Its new rôle was as a fuel bunker for plant machinery, and it was towed from site to site by means of the A-frame attached to its front. A semaphore trafficator is fixed to the cab, and a hand-operated pump and fuel-line are mounted on the body

side. With its lofty bonnet and shapely tanker body it would have been an ideal candidate for preservation.

Right: Registered 2961 R, this LAD-cabbed Leyland Octopus still retained the original operator's paintwork on the chassis and cab. The cab featured a neat step between the bumper and front wheel, and this, coupled with the deep door, gave good access to what appeared to be a high cab. Illustrated at Nottingham Goose Fair on 27 September 1981, it was owned by Holland's Amusements, and was travelling with that firm's Cake Walk ride. By this time the original bodywork had been replaced by a large box-van body.

In the war years the USA built large numbers of 6-ton 6x6 vehicles as heavy artillery tractors, many being supplied to other Allied countries through the Lend-Lease programme. They were built by a variety of manufacturers, but one of the most well-known was the Mack NM6, which was a development of a commercial model. Originally powered by a 159bhp six-cylinder engine, this proved to be a very durable machine, and after the war many continued in service with various governments before finding their way onto the civilian market.

In the UK the NM6 became a regular sight on fairgrounds and with circus operators, one of the most famous being Chipperfield's, which employed a large number to move its vast tenting operation around the country. Parked adjacent to the entrance to the Big Top in Nottingham on 4 April 1964 are three of the fleet, all fitted with custom-built bodies behind the original cabs. Seen here, the cabs have simple metal roofs, but these would have been canvas in wartime guise.

It was common practice to park the heavy transport vehicles around the perimeter of the circus site, forming a boundary and acting as an advert with the name prominently painted on the body sides.

Left: Perhaps the most famous vehicle in the Chipperfield's fleet was the Mack NM6 crane truck, immortalised by the Corgi die-cast model which is now a sought-after collector's item.

The late 1949/early 1950-registered original was caught in Nottingham on 4 April 1964, the bunting on its jib blowing strongly in the cool spring breeze. The cab, with no side windows, must have been a very draughty place to work, conditions unlikely to be tolerated today.

A postwar Albion stands to the left of the picture.

Above: Another example of the Mack NM6 is seen in Nottingham for the Goose Fair on 4 October 1964. JLV 760 was owned by J. P. Collins, and is seen here whilst manoeuvring his living-van by means of the front-mounted coupling pin. This vehicle, used to move and provide power for his Waltzer ride, is basically very original, although the cab has been roofed. The original body remains, as does the winch mounted behind the cab, whilst above it are fuel drums for the diesel generators fitted within the body. The caravan is in classic showman's style and is very substantially built.

The major producer of 7½-ton 6x6 trucks in the USA for the duration of the war was Mack with its NO model, popularly known as the 'Super Mack'. This was powered by the manufacturer's own 'EY' 159bhp six-cylinder engine, which drove through a five-speed gearbox to all wheels. It was noted for its exceptional ground clearance, provided by the double-reduction gearing fitted at the ends of the front axle, dispensing with the need for a conventional universal joint. Fitted with a front-mounted Garwood winch, this was a very useful machine, and was used extensively to pull the 155mm 'Long Tom' field gun.

The example seen here makes a very impressive breakdown vehicle, and is towing a preserved Glasgow double-deck trolleybus near Wakefield on 12 July 1975. It was based at Hatfield, near Doncaster.

Right: Merryweather of Greenwich, London, is one of the most famous names in the history of fire-fighting, with a pedigree going back many years. Once it ceased building steam appliances and moved on to petrol-driven machines, many were built using the firm's own name including E 2152, an A-type fire pump of 1913. Originally supplied on solid tyres to Bass, Ratcliffe & Gretton Ltd, for use at that company's Burton-on-Trent brewery, it was fitted with a 53hp four-cylinder Aster engine and a Hatfield 350/400gpm pump. This was rumoured once to have pumped nonstop for three days and nights at a grain-store fire in the 1950s.

   The vehicle is seen here after withdrawal in a scrapyard at Langley Mill on 17 August 1963. Still wearing its World War 2 white paint, and with most of its small items of equipment missing, it presents a sorry sight, although the splendid original headlamps remain. Fortunately it was soon acquired for preservation. All its companions in the yard would now be suitable candidates for preservation, although in the early 1960s these more modern vehicles did not command the same interest, and most were lost forever.

Right: During the mid-1930s Morris Commercial used the slogan 'British to the backbone', and advertised itself as 'The largest British manufacturers specialising exclusively in commercial vehicles'. The 4-ton Leader model was the largest in its goods vehicle range, being built in both short- and long-wheelbase forms, and available with either a four- or six-cylinder engine. However, it never sold in such large numbers as the lighter models in the range.
This very original-looking breakdown lorry with chariot body was parked behind a garage in Stamford, Lincolnshire, on 12 April 1964. An example of Austin's A40 10cwt pick-up dating from the 1950s can be seen in the background.

Waiting for its full complement of passengers, for a journey out across the sand to the waiting pleasure boat at Blackpool on 20 September 1980, is this wartime Morris Commercial C8. Built as an artillery tractor, for use with a two-pounder gun, it is powered by a 70bhp four-cylinder engine driving all the wheels through a five-speed gearbox, and would originally have carried a crew of four plus the driver. Now brightly painted, it has had its exhaust system rerouted to keep it well above water level for when it wades out to boats.

The 15cwt Morris Commercial CS8 was one of a large number of similar trucks made by all the major manufacturers and widely used by all branches of the armed forces. Powered by a 60bhp six-cylinder engine that drove the rear wheels only, the Morris was one of the earlier designs, and many were lost on the French beaches during the 1940 retreat.

This particular example has retained its original GS (General Service) body, and the original open driving compartment has been enclosed by a well-built cab. It is seen at the Perkins Diesel Engines sports day on 22 June 1963, keeping company with the Leyland Cub illustrated earlier.

The 10cwt Morris J-type van was unveiled to the public at the Commercial Motor Show in October 1948, with production starting in September 1949. As introduced it was powered by a 36bhp 1,476cc side-valve four-cylinder petrol engine, driving through a three-speed gearbox. It was designed as a small commercial vehicle from the start, rather than being based on a car design, and the standard works-built van had an integral all-steel body with sliding cab doors. In February 1957 a larger OHV engine was fitted and the gearbox changed to a four-speed unit, the revised model being known as the JB.

This very smart 1951 example, seen at Bressingham Gardens and Steam Museum on 8 June 1978, carries a stylish ice cream sales body, and still wears a BMC Drivers' Club badge on its radiator grille.

At one time it was not unknown for a business to run a Rolls-Royce converted into van form. Usually it was done for the publicity that such a vehicle would attract, but the legendary longevity of the engine and chassis components must have had some bearing on any decision to run such a vehicle, entailing as it did a substantial capital outlay. Typical operators of such a machine would be high-class clothing or ladies' dress manufacturers — the equivalent of today's 'designer labels' — or sometimes up-market confectioners or florists, catering for the sort of *clientele* who would welcome a home delivery by a vehicle as auspicious as a Rolls-Royce.

WD 9257, a 1935 model, was owned by Ford & Weston Ltd, building contractors of Derby and Cheltenham, and certainly provided a good advertisement with its classic lines, smart coachwork and smiling lady driver. It was seen on the borders of Derbyshire and Leicestershire on 31 May 1964, acting as a sound equipment van at a gala day.

Left: Scammell continued to produce chain-drive models long after most European manufacturers had ceased to do so. The example shown here may have been part of Scammell's 1946 production run of 45-ton chain-drive tractors. Fitted with a Gardner 6LW engine and a breakdown crane, it had recently been withdrawn from service by a Nottingham haulage contractor, being in a north Nottinghamshire scrapyard on 23 May 1963. Although looking well-used, it was in running order and was, the writer recalls, for sale at £250.

At the left of the picture is an early postwar Leyland Beaver, piled high with the remains of other vehicles, whilst on the right is an elderly tractor loader, few of which appear to have survived into the preservation era.

Below: Scammell was renowned for building specialist vehicles, and immediately after the war introduced a vehicle for a very specialist market.

This was the 'Showtrac', designed to meet the needs of the travelling showman and based on the 20-ton ballast tractor. Built on a 10ft wheelbase, and with a 450A Mawdsley dynamo mounted on a ballast block in the rear of the body, it was coachbuilt by Brown Bros of Tottenham, and presented a squat, powerful appearance. The dynamo was driven from the gearbox by a shaft and multiple vee belt. Power was provided by a Gardner 6LW engine, although the last two to be built had six-cylinder Meadows engines. Only 18 true Showtracs were built, between December 1945 and September 1948, although subsequent conversions were produced both by Scammell and by individual operators.

Pulling on to the Forest for Nottingham's Goose Fair on 30 September 1963 is HTO 221. Chassis No 6114, it was delivered new to Hibble & Mellors of Nottingham on 1 June 1946. It is seen hauling a magnificently-painted, traditionally-styled showman's box-van trailer.

Left: Evidence that the Shell-Mex Scammell Highwayman tractors were well cared for is provided by this example, 220 BGO. Converted to a showman's draw-bar tractor and fitted with generators, it was at Nottingham Goose Fair on 10 October 1982. After 22 years' hard life on the road, much of it spent pulling tankers, it still looked good — even its bumpers shone. It was owned by A. W. Price, and was being used with a Paratrooper ride.

To the left is the front of a glass-fibre-cabbed Foden S80, with its characteristically large headlamps.

Right: With the end of the war the Army required a suitable replacement for the 6x4 Pioneer which had been used both as a recovery tractor and as the motive power for tank transporter trailers. This need dictated a 6x6 recovery machine and, to fit the bill, Scammell produced the Explorer, as the Army's standard 10-ton tractor. Over 1,000 were built between 1948 and 1953 for the British and Commonwealth armies. Most were built with Scammell/Meadows 10.35-litre petrol engines which, driving through a six-speed constant-mesh gearbox, were somewhat faster than the pedestrian Pioneer. The cab was double-skinned and insulated for use in varying climatic conditions.

HBK 598N was at Wakefield on 26 March 1977, travelling with the Austen Bros circus.

Left: For towing 20-ton, low-loading trailers the Army required a 6x6 tractor, and to fill this need Scammell produced an army version of the Constructor, designated FV12101. The body space was divided into compartments to carry ballast, equipment and tools, and power was provided by a Scammell/Meadows engine driving through a six-speed gearbox.

81 BL 97 is seen in Derbyshire, coupled to a typical drawbar trailer, on 30 July 1977.

Above: The Birmingham & Midland Motor Omnibus Co Ltd or 'Midland Red', as it was more commonly known, had a very charismatic and inventive Chief Engineer, Mr Wyndham Shire, who, from 1923 until 1940, designed many of the company's vehicles, these being known by the letters SOS. There appears to be no precise definition of this (although several ideas exist), but the writer grew up understanding that it stood for 'Shire's Own Specification'.

Some chassis were sold to other operators, and RC 7923 seen here is an ex-Trent SON model, dating from 1939, with bodywork by Willowbrook. The 'TMT' cast into the top of the radiator stands for Trent Motor Traction. Owned by a Wigston, Leicestershire showman it was found at Basford Wakes (fair), Nottingham, on 16 August 1964. The exhaust pipe has been extended up the body side to carry the fumes away while it was generating power for lights.

Above: Thornycroft of Basingstoke was a long-lived quality vehicle manufacturer, making a wide range of goods vehicles right up to the Mighty Antar tank-transporter tractor at the top of the range. The company was taken over by AEC in December 1960, and production of normal load-carrying lorries ceased within a year.

During the 1940s and '50s Thornycroft competed quite strongly in the medium load range of 6-8 tons, and here we see a 1959 6-ton Swiftsure, a model introduced in 1957. This handsome, large-capacity van was new to a Nottingham operator, and is seen in its home city on 10 October 1982, attending the Goose Fair with a subsequent owner. By this time, Thornycrofts in regular use had become quite rare, the make having been out of production for approximately 20 years.

Right: Tilling-Stevens of Maidstone, Kent, was renowned for its persistence with, and perfection of, petrol-electric transmission, whereby the vehicle's engine drove a dynamo, providing power to an electric motor which in turn drove the road wheels. The system, working at 110V, was flexible, completely variable and removed the necessity for gear-changing. This was an important factor in the years immediately prior to World War 1, when both horsebus and tram drivers had to be taught the niceties of manual gearboxes.

CD 3806 is a 1912 TS3 vehicle that was originally a Brighton area bus. It was bought by a showman, George Pickard, in 1926, and fitted with a wooden van body. Converted to run on pneumatic tyres in the 1930s, it regularly attended many of the well-known London fairs. Seen at the Great Steam Fair at White Waltham on 30 August 1964, it was generating electricity with its immaculate engine. It is rumoured to be still in existence.

Left: Once in a while in transport history a manufacturer comes along and applies some lateral thinking to a problem, coming up with an original, but thoroughly workable, idea. One such manufacturer was Tilling-Stevens, which built some searchlight lorries for the Army between 1936 and 1945. The company's petrol-electric system was ideal for this purpose, a 70bhp four-cylinder Tilling petrol engine, mounted in the cab, driving a dynamo under the bonnet where the engine would normally be found. To assist with cooling, as the system was run for long periods with the vehicle stationary, twin radiators were fitted within the snub nose.

Derby Corporation 4, ARC 267, was a 3-ton TS20 model used as a tower wagon for trolleybus overhead line maintenance, and is seen at work in the city centre on 13 September 1964. It will be noted that it has no headlights, only side lamps and a spotlight.

Front cover: TV 7597 was a 1932 Bedford WS 30cwt lorry, first registered in January 1933 (chassis number 83513; engine number 426058). Whilst resembling a preserved vehicle with its elaborate and high standard of finish it was actually very much a working machine. It was owned from new by a Mr Simmons of Nottingham, a glass and china factor, and was used to collect stock direct from the Staffordshire potteries for use in his business.

The vehicle was lovingly cared for by the owner and his son, and an extra hour was always added to journey times to allow for leathering it down on a wet day — it was never put into its garage wet. When seen it still had the original engine block, which had been rebored twice, and the writer was assured that it had 'been round the clock' at least once. A Bedford Drivers Club badge can just be discerned on the bumper.

It was regularly seen on the roads around the centre of Nottingham in the 1960s, being photographed on the Victoria Embankment, at the side of the River Trent, on 3 October 1965.

Back cover: To many enthusiasts the Scammell Highwayman epitomises the marque, with its impressive stance and the classical lines of the radiator and bonnet. Disliked by some drivers due to its tricky gate-change gearbox, it was nonetheless loved by operators, possibly due to its undoubted strength and an ability to 'soldier on'. 226 BGO is a well-turned-out 1960 example, owned by Shell-Mex and BP Ltd, which distributed fuel for both companies. Coupled to a frameless tanker trailer, it is seen basking in the sun at a Lorry Driver of the Year heat at Melton Mowbray, Leicestershire, on 21 July 1963. Its paintwork shines, the chromed twin front bumper gleams and the whole vehicle is a credit to its owners.